Compilation copyright © 1990 by Dick Bruna Books, Inc.
Illustrations Dick Bruna, copyright © Mercis bv, 1959, 1962, 1964, 1966, 1967, 1972, 1973, 1974, 1982, 1984, 1986

Created and manufactured by Dick Bruna Books, Inc., by arrangement with Ottenheimer Publishers, Inc. Illustrations by Dick Bruna.
No part of this book may be reproduced in any form without written permission from the publisher.

First published in Great Britain in 1990 by William Collins Sons & Co Ltd,
8 Grafton Street, London W1X 3LA

A CIP catalogue record for this book is available from the British Library

0 00 184579 9

Printed in Italy

I know my abc
Dick Bruna

COLLINS

Aa

apple

Ff

frogs

Gg

guitar

Hh

horse

Ii

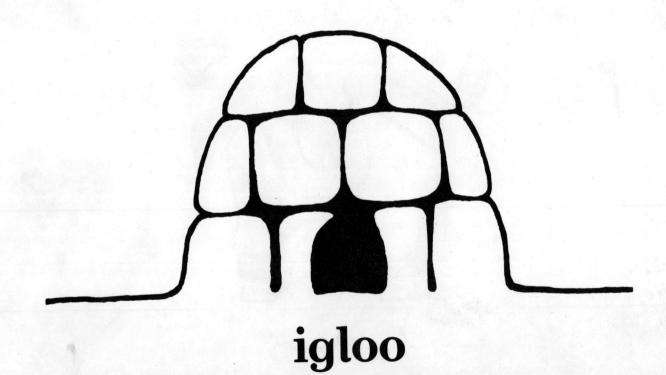

igloo

Jj

jacket

Kk

koala

lion

Ll

mouse

Mm

Nn

necklace

Oo

octopus

P p

pear

Qq

queen

Rr

rain

Ss

strawberry

Tt

teddy

Uu

umbrella

Vv

violin

Ww

walrus

Xx

xylophone

Yy

yacht